GAO

I0470688

October 2012

ELECTRICITY

Significant Changes Are Expected in Coal-Fueled Generation, but Coal is Likely to Remain a Key Fuel Source

GAO
Accountability ★ Integrity ★ Reliability

GAO-13-72

Highlights of GAO-13-72, a report to the Chairman, Committee on Commerce, Science and Transportation, U.S. Senate

ELECTRICITY

Significant Changes Are Expected in Coal-Fueled Generation, but Coal Is Likely to Remain a Key Fuel Source

Why GAO Did This Study

Coal is a key domestic fuel source and an important contributor to the U.S. economy. Most coal produced in the United States is used to generate electricity. In 2011, 1,387 coal-fueled electricity generating units produced about 42 percent of the nation's electricity. After decades of growth, U.S. coal production and consumption have fallen, primarily due to declines in the use of coal to generate electricity.

According to the Environmental Protection Agency (EPA), using coal to generate electricity is associated with health and environmental concerns such as emissions of sulfur dioxide, a pollutant linked to respiratory illnesses, and carbon dioxide, a greenhouse gas linked to climate change. In response to recent environmental regulations and changing market conditions, such as the recent decrease in the price of natural gas, power companies may retire some units, which could affect the coal fleet's generating capacity— the ability to generate electricity—and the amount of electricity generated from coal. Power companies may also retrofit some units by installing controls to reduce pollutants.

GAO was asked to examine (1) how the fleet of coal-fueled electricity generating units may change in the future in terms of its generating capacity and other aspects and (2) the future use of coal to generate electricity in the United States and key factors that could affect it. GAO conducted a statistical analysis of plans for retiring coal-fueled units, interviewed stakeholders, and reviewed information on industry plans and long-term forecasts by EIA and others. GAO is not making any recommendations in this report.

View GAO-13-72. For more information, contact Frank Rusco at (202) 512-3841 or ruscof@gao.gov.

What GAO Found

Retirements of older units, retrofits of existing units with pollution controls, and the construction of some new coal-fueled units are expected to significantly change the coal-fueled electricity generating fleet, making it capable of emitting lower levels of pollutants than the current fleet but reducing its future electricity generating capacity. Two broad trends are affecting power companies' decisions related to coal-fueled generating units—recent environmental regulations and changing market conditions, such as the recent decrease in the price of natural gas. Regarding retirements, forecasts GAO reviewed based on current policies project that power companies may retire 15 to 24 percent of coal-fueled generating capacity by 2035—an amount consistent with GAO's analysis. GAO's statistical analysis, examining data on power companies that have announced plans to retire coal-fueled units, found that these power companies are more likely to retire units that are older, smaller, and more polluting. For example, the units companies plan to retire emitted an average of twice as much sulfur dioxide per unit of fuel used in 2011 as units that companies do not plan to retire. Based on the characteristics of the units companies plan to retire, GAO estimated additional capacity that may retire. In total, GAO identified 15 to 18 percent of coal-fueled capacity that power companies either plan to retire or that GAO estimated may retire—an amount consistent with the forecasts GAO reviewed. Regarding retrofits, the coal-fueled generating fleet may also become less polluting in the future as power companies install controls on many remaining units. Regarding new coal-fueled units, these are likely to be less polluting as they must incorporate advanced technologies to reduce emissions of regulated pollutants. Coal-fueled capacity may decline in the future as less capacity is expected to be built than is expected to retire.

According to stakeholders and three long-term forecasts GAO reviewed, coal is generally expected to remain a key fuel source for U.S. electricity generation in the future, but coal's share as a source of electricity may continue to decline. For example, in its forecast based on current policies, the Energy Information Administration (EIA) forecasts that the amount of electricity generated using coal is expected to remain relatively constant through 2035, but it forecasts that the share of coal-fueled electricity generation will decline from 42 percent in 2011 to 38 percent in 2035. Available information suggests that the future U.S. use of coal may be determined by several key factors, including the price of natural gas and environmental regulations. For example, available information suggests that the price of coal compared with other fuel sources will influence how economically attractive it is to use coal to generate electricity. EIA assessed several scenarios of future fuel prices and forecasts that coal's share of U.S. electricity generation will fall to 30 percent in 2035 if natural gas prices are low or 40 percent if natural gas prices are high. In addition, some stakeholders told GAO that the future use of coal could be significantly affected if existing environmental regulations become more stringent or if additional environmental regulations are issued. For example, EIA forecasts that two hypothetical future policies that reduce carbon dioxide emissions from the electricity sector by 46 percent and 76 percent would result in coal's share of U.S. electricity generation falling to 16 and 4 percent in 2035, respectively.

EPA provided technical comments that were incorporated as appropriate.

_____ **United States Government Accountability Office**

Contents

Figures

Abbreviations

Btu	British thermal unit
CO_2	carbon dioxide
EIA	Energy Information Administration
EPA	Environmental Protection Agency
IEA	International Energy Agency
MW	megawatt
MWh	megawatt-hour
NO_x	nitrogen oxides
SO_2	sulfur dioxide

October 29, 2012

The Honorable John D. Rockefeller IV
Chairman
Committee on Commerce, Science, and Transportation
United States Senate

Dear Mr. Chairman:

Coal is a key domestic fuel source and an important contributor to the U.S. economy. The United States has the largest recoverable coal reserves in the world, according to the Energy Information Administration (EIA).[1] In 2011, about 86,200 workers around the country produced more than 1 billion tons of coal, and more than 90 percent of this coal was used to generate electricity in the United States.[2] Also in 2011, there were 1,387 coal-fueled electricity generating units with a total of 317,469 megawatts (MW) of capacity—a measure of the ability to generate electricity[3]—about 30 percent of total generating capacity in the United States. These coal-fueled units generated 42 percent of the nation's electricity in 2011.[4] After decades of growth—peaking in 2008—U.S. coal production has fallen, primarily due to declines in the use of coal to generate electricity. Some stakeholders expressed concern, if this trend were to continue, about the implications on electricity systems,

---

[1]EIA is a statistical agency within the Department of Energy that collects, analyzes, and disseminates independent information on energy issues.

[2]The rest of production was exported or used for other purposes, including steel production. In addition to coal, electricity is produced using other fossil fuels, particularly natural gas; through nuclear fission; and using renewable sources, including hydropower, wind, geothermal, and solar energy.

[3]Generating capacity is measured in MW and refers to the maximum capability of a unit to produce electricity. A unit with 1,000 MW of capacity can generate up to 1,000 megawatt-hours (MWh) of electricity in 1 hour, enough to provide electricity for up to 1 million homes. There are many measures of capacity, and we generally refer to net summer capacity in this report—a generating unit's capacity to produce electricity during the summer when electricity demand for many electricity systems and losses in efficiency are generally the highest. Net capacity data excludes output used internally for plant operations.

[4]Electricity generation depends on the capacity of generating units and how often and to what extent units are operated.

communities, and economies that rely on coal mining and coal-fueled electricity.

Using coal to generate electricity has been associated with human health and environmental concerns by the Environmental Protection Agency (EPA), the primary federal agency responsible for implementing many of the nation's environmental laws. For example, according to EPA data, coal-fueled electricity generating units are among the largest emitters of sulfur dioxide (SO_2) and nitrogen oxides (NO_x), which have been linked to respiratory illnesses and acid rain. EPA recently proposed or finalized several regulations, as required or authorized, that aim to address certain health or environmental impacts associated with coal-fueled electricity generating units. In response to these regulations, power companies might retrofit some units by installing controls to reduce pollutants or by taking other steps to reduce adverse impacts.[5] When it is not economic to take these actions, power companies may retire some units, which could affect coal-fueled generating capacity and the amount of electricity actually generated from coal. We recently reported on the price and reliability implications of key recent EPA regulations.[6] In addition, according to EPA data, coal-fueled electricity generating units emit large quantities of carbon dioxide (CO_2). As we have previously reported, compared with natural gas-fueled units, coal-fueled units produced, on average, over twice as much CO_2 per unit of electricity produced as natural gas units in 2010.[7] The National Research Council[8] has stated that emissions of CO_2 and other greenhouse gases are linked to climate change. There have been a number of legislative proposals in Congress, regulatory action by EPA, and actions at the state and local levels aiming to reduce CO_2 emissions.

---

[5]Compliance with regulations may involve using various technologies or making infrastructure changes to reduce adverse impacts; for example, installing liners at facilities used to store coal combustion wastes to minimize leaching of contaminants into groundwater.

[6]GAO, *EPA Regulations and Electricity: Better Monitoring by Agencies Could Strengthen Efforts to Address Potential Challenges,* GAO-12-635 (Washington, D.C.: July 17, 2012).

[7]See GAO, *Air Emissions and Electricity Generation at U.S. Power Plants,* GAO-12-545R (Washington, D.C.: Apr. 18, 2012).

[8]The National Research Council is the principal operating agency of both the National Academy of Sciences and the National Academy of Engineering.

You asked us to examine the future use of coal to generate electricity. Our objectives for this report were to examine what available information indicates about: (1) how the nation's fleet of coal-fueled electricity generating units may change in the future in terms of its generating capacity and other aspects and (2) the use of coal to generate electricity in the United States in the future and key factors that could affect it.

To examine how the coal-fueled generating fleet may change, we used data from Ventyx Velocity Suite, a proprietary database containing consolidated energy and emissions data from EIA, EPA, and other sources. We used data as of July 27, 2012, to describe characteristics of coal-fueled electricity generating units and to provide information on power companies' plans to retire coal-fueled units and build new ones. Such information reflects publicly reported plans as identified by Ventyx. As plans may change, the actual number and characteristics of future retirements and new construction of coal-fueled units may differ from what is represented in Ventyx as of July 2012. To assess the types of units that may be retired, we carried out a statistical analysis of units owned by power companies that have announced plans to retire coal-fueled units. We analyzed various characteristics including characteristics of the unit (i.e., size and age) and the characteristics of the power company that owns the unit (i.e., whether it is traditionally regulated or operates in a restructured market).[9] We then examined units owned by companies that have not announced any planned retirements in order to estimate how many of those units companies may consider retiring. Our statistical analysis did not examine the amount of electricity that may be generated at coal-fueled units in the future. Appendix I provides further information about our statistical analysis. To provide information about the use of coal to generate electricity in the future and key factors that could affect it, we reviewed forecasts from EIA, the International Energy Agency

---

[9]In some areas of the country, referred to as "traditionally regulated markets," state public utility commissions—which generally aim to ensure retail electricity rates are just and reasonable—review power companies' requests to recover the costs of investments in new generating units, distribution lines, and other system upgrades. Once a state public utility commission approves a power company's request, consumer retail prices are adjusted to recover the power company's costs plus a rate of return. In other areas of the country, referred to as "restructured markets," electricity is sold by multiple companies competing with each other. In these areas, public utility commissions play a more limited role in overseeing generation. Consumers pay retail electricity rates based on the price of electricity as determined in wholesale markets.

(IEA)[10] and IHS Global Insight,[11] and summarized projections of coal-fueled electricity generation under different scenarios. Appendix II summarizes the scenarios we examined. While long-term forecasts are subject to inherent uncertainties, we found the EIA, IEA and IHS Global Insight forecasts to be reasonable for describing what is known about the potential future use of coal to generate electricity. To respond to both objectives, we reviewed available literature, including studies by federal agencies and research organizations, and summarized the results of semistructured interviews with a nonprobability sample of 36 stakeholders. Stakeholders included representatives from power companies, a coal company, and nongovernmental organizations, and officials from federal and state agencies. We selected these stakeholders to be broadly representative of differing perspectives on these issues based on recommendations from agencies and industry associations, along with other information. Because we used a nonprobability sample, the views of these stakeholders are not generalizable to all potential stakeholders, but they provide illustrative examples of views. To provide information on recent electricity industry trends, we summarized historical data from EIA. To assess the reliability of Ventyx and EIA historical data, we reviewed existing documentation, interviewed Ventyx and EIA staff, consulted with agency officials and other knowledgeable parties, conducted some electronic testing, and compared data in Ventyx to information obtained from several power companies and regional transmission organizations. We determined the Ventyx and EIA data to be sufficiently reliable for the purposes of this report. Some numbers in this report have been rounded.

We conducted our work from July 2011 to October 2012 in accordance with generally accepted government auditing standards. Those standards require that we plan and perform the audit to obtain sufficient, appropriate evidence to provide a reasonable basis for our findings and conclusions based on our audit objectives. We believe that the evidence obtained provides a reasonable basis for our findings and conclusions based on our audit objectives.

---

[10]IEA is an international organization composed of 28 of the member nations of the Organisation for Economic Co-operation and Development that, among other things, collects energy data and provides research and analysis on ways to ensure reliable, affordable, and clean energy.

[11]IHS Global Insight is a firm that provides comprehensive economic and financial information on countries, regions, and industries.

Background

Because of its abundance and historically low cost, coal is an important fuel source in the United States, accounting for about 20 percent of total energy use in 2011. Nearly all coal consumed in the United States is produced domestically, and coal represents about 29 percent of all domestically produced energy. U.S. coal production generally increased since 1960 and reached its highest level in 2008. Advancements in mining technology and a shift to using surface mines to a greater extent than underground mines has boosted coal's overall productivity and enabled production to increase even as the number of workers decreased. In 2011, half as many workers produced 24 percent more coal than in 1985, as shown in figure 1. Data from the Bureau of Labor Statistics indicate that about 86,200 people were employed in coal mining in the United States in 2011.

Figure 1: Coal Production and Employment, 1960-2011

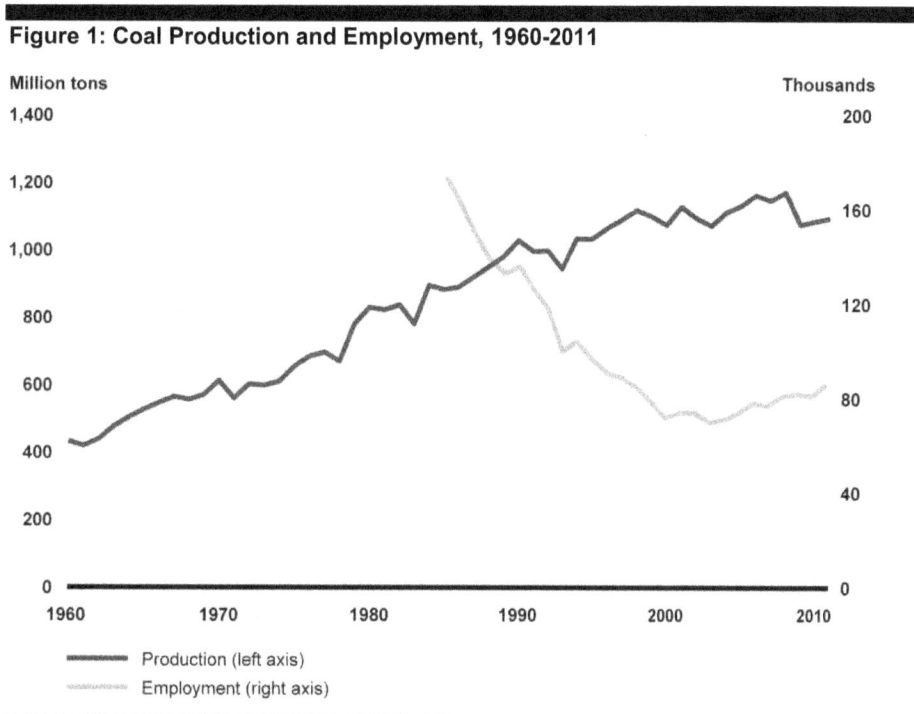

Production (left axis)
Employment (right axis)

Sources: GAO analysis of EIA and Bureau of Labor Statistics data.

In the United States, coal is primarily used to generate electricity—over 90 percent of coal was used to generate about 42 percent of electricity in 2011. The amount of electricity generated using coal has generally increased since the 1960s, but decreased recently due to a combination of a decline in overall electricity demand, shifts in the relative prices of fuels, and other reasons. (See fig. 2.) Meanwhile, coal's share of total electricity generation has fluctuated over time. EIA has stated that several factors, including low oil prices during the late 1960s—which served to increase electricity generation from oil—and the oil price shocks of the 1970s have influenced the mix of fuel sources used to produce electricity.

Figure 2: Electricity Generation from Coal, 1960-2010

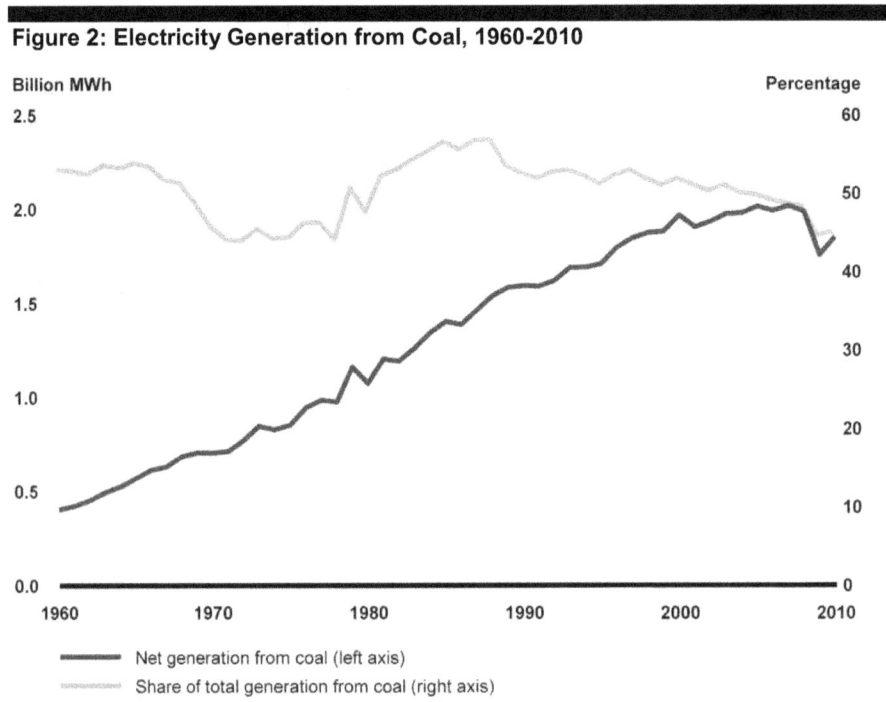

Net generation from coal (left axis)
Share of total generation from coal (right axis)

Source: GAO analysis of EIA data.

Note: Net generation excludes electricity generation that is used internally for plant operations.

Two broad trends—recent environmental regulations and changing market conditions—are affecting power companies' decisions related to coal-fueled electricity generating units. Regarding environmental regulations, as we have previously reported, since June 2010, EPA proposed or finalized several regulations that would reduce certain adverse health or environmental impacts, including impacts associated with coal-fueled electricity generating units.[12] These regulations have potentially significant implications for public health and the environment. One of the most significant regulations in terms of EPA's estimated benefits and costs, EPA's Mercury and Air Toxics Standards, establishes emissions limitations on mercury and other toxic pollutants. Mercury is a toxic element, and human intake of mercury, for example, through consumption of fish that ingested the mercury, has been linked to a wide range of health ailments. In particular, mercury can harm fetuses and cause neurological disorders in children, resulting in, among other things, impaired cognitive abilities. Other toxic metals emitted from power plants, such as arsenic, chromium, and nickel can cause cancer. EPA estimates that its finalized regulation would reduce mercury emissions from coal-fueled electricity generating units by 75 percent, as well as reduce SO_2 and other emissions. EPA estimated the benefits of this one regulation would be $39 to $96 billion with costs of $10.2 billion in 2016 (in 2011 year dollars). The requirements and deadlines these regulations may establish for generating units are uncertain. In particular, several regulations have not been finalized, and finalized regulations could be subject to legal challenges that result in changes. For example, EPA finalized the Cross-State Air Pollution Rule in August 2011. The regulation would have required reductions of certain emissions of air pollutants in 28 states because some of these pollutants may travel in the atmosphere and impact air quality in other states. The U.S. Court of Appeals for the D.C. Circuit recently struck down the regulation, and EPA has asked the full court to rehear the case, creating uncertainty as to what may be required from generating units in the future to address such

---

[12]See GAO-12-635. Specifically, these include the Cross-State Air Pollution Rule, which would have prohibited certain emissions of air pollutants in 28 states because of the impact they would have on air quality in other states; the National Emissions Standards for Hazardous Air Pollutants from Coal and Oil Fired Electric Utility Steam Generating Units, also known as the Mercury and Air Toxics Standards, which establishes emissions limitations on mercury and other toxic pollutants; the proposed Cooling Water Intake Structures at Existing Facilities and Phase I Facilities regulation, which would establish requirements for water withdrawn and used for cooling purposes that reflect the best technology available to minimize adverse environmental impact; and the proposed Disposal of Coal Combustion Residuals from Electric Utilities regulation, which would govern the disposal of coal combustion residuals, such as coal ash, in landfills or surface impoundments. On April 13, 2012, EPA also proposed new source performance standards for greenhouse gas emissions from certain new fossil fuel electricity generating units—including coal-fueled units—but the standards would not apply to existing units.

emissions.[13] In response to these regulations, power companies might retrofit generating units with controls to reduce pollutants and, when it is not economic to retrofit, may retire some generating units.

Regarding broader market conditions, important market drivers have been weighing on the viability of coal-fueled electricity generating units. Key among these has been the recent decrease in the price of natural gas, which has made it more attractive for power companies to build new gas-fueled electricity generating units and to utilize existing units more. In addition, slow expected growth in demand for electricity in some areas has decreased the need for new generating units. Power companies may weigh the costs of any needed investments compared with the benefits of continuing to generate electricity at a particular unit. When the costs outweigh the benefits, a power company may decide to retire a unit rather than continue to operate the unit or install new pollution control equipment.

The majority of coal produced in the United States is used domestically, though exports represent a small but recently growing fraction of U.S. coal production. In 2010, the United States exported 82 million tons of coal, which accounted for 8 percent of total production. As shown in figure 3, coal exports to European and Asian markets represented 76 percent of total U.S. coal exports in 2011. In 2011, total coal exports were up 31 percent compared with 2010, reaching 107 million tons, due largely to rising exports to Europe and Asia. This was the highest level of exports since 1991. In 2011, 35 percent of U.S. coal exports were of the types of coal typically used to produce electricity, the remainder were of metallurgical coals used in industrial processes, such as steelmaking.

---

[13]Specifically, the court issued an opinion that would strike down the regulation but has not issued an order striking it down and likely will not issue such an order before deciding whether the full court will rehear the case.

Figure 3: U.S. Coal Exports by Destination, 2001-2011

Million tons

Legend:
- Africa, Australia, and Oceania
- South America
- North America
- Asia
- Europe

Source: EIA.

To better understand the potential future of the coal and electricity industries, the federal government, private companies, and others use models to project future industry conditions, including the future use of coal. For example, EIA, IEA, and IHS Global Insight produce long-term projections of electricity generation and generation from coal.[14] Because the future depends on a multitude of factors that are difficult to predict, EIA assesses various scenarios with different assumptions about future conditions to better understand the range of potential future outcomes. For example, EIA's primary scenario, called its "reference" scenario, is a business-as-usual estimate based on existing policies, known technology,

---

[14]See: EIA, *Annual Energy Outlook 2012,* DOE/EIA-0383 (Washington, D.C.: June 2012); IEA, *World Energy Outlook 2011* (Paris, France: 2011); and IHS Global Insight, *U.S. Energy Outlook*, September 2011.

and current technological and demographic trends. Additional scenarios make different assumptions about fuel prices, economic conditions, and government policies, among other things. Some of these scenarios are especially relevant to the question of coal's future, because they address factors currently affecting the industry, such as the prices of coal and natural gas—a fuel that competes with coal—and possible future policies to address climate change. Appendix II presents further information about the major assumptions behind these forecasts and scenarios.

Retirements, Retrofits, and New Construction May Result in a Smaller but Cleaner Coal-Fueled Electricity Generating Fleet

The nation's fleet of coal-fueled electricity generating units may have less total generating capacity in the future, and the fleet may be capable of emitting lower levels of pollutants, according to available information. These changes will be driven by industry plans to retire a significant number of units, install pollution control equipment on others, and build a few, new coal-fueled units that may emit lower levels of pollutants than the current fleet's average emissions.

Power Companies Are Planning to Retire a Significant Number of Older, Smaller, More Polluting Units

According to forecasts we reviewed, power companies may retire a significant number of coal-fueled units in the future. In its reference scenario reflecting current policies, EIA projects that power companies may retire 49,000 MW of coal-fueled capacity from 2011 through 2035 (i.e., 15 percent of coal-fueled capacity in 2011). IHS Global Insight projects that power companies may retire 76,476 MW of capacity from 2011 through 2035 (i.e., 24 percent of coal-fueled capacity in 2011).

Our statistical analysis of Ventyx data on announced retirement plans indicates that, among other things, companies are planning to retire units that are older, smaller, and more polluting. To assess the types of units that may be retired, we analyzed data on current power company plans to retire coal-fueled units. According to Ventyx data, power companies have already reported plans to retire 174 coal-fueled units with a total 30,447 MW net summer capacity through 2020—which accounted for 10 percent of coal-fueled capacity in 2011.[15] As we have previously reported, this

---

[15]Data presented throughout this section refer to units with over 25 MW of net summer capacity.

would be significantly more retirements than have occurred in the past—
almost twice as much coal-fueled capacity as retired in the 22 years from
1990 through April 2012.[16] Based on our statistical analysis of these
plans, power companies are more likely to plan to retire units that are
older, smaller, and more polluting. (Appendix I provides further
information on our statistical analysis, which included examining several
other characteristics that may affect plans to retire units such as
(1) whether power companies are traditionally regulated or operate in
restructured markets and (2) a unit's cost of generating electricity relative
to regional prices.)

- *Older.* Power companies' plans indicate they are more likely to retire
  older coal-fueled electricity generating units than newer units. Today's
  fleet of operating coal-fueled units was built from 1943 through 2012,
  with the bulk of the capacity built in the 1970s and early 1980s. As
  shown in figure 4, units that power companies plan to retire are
  generally older, on average 54 years old compared with units with no
  retirement plans that average 39 years old. Some stakeholders we
  interviewed said that power companies are more likely to retire older
  units because these units may be reaching the end of their useful
  lives, can be less efficient at converting coal to electricity, and can be
  more expensive than newer units to retrofit, maintain, and operate.

---

[16]GAO-12-635.

Figure 4: Capacity of Coal-Fueled Units by Year of First Commercial Operation and Planned Retirements

Thousand MW

Retirement planned

No retirement planned

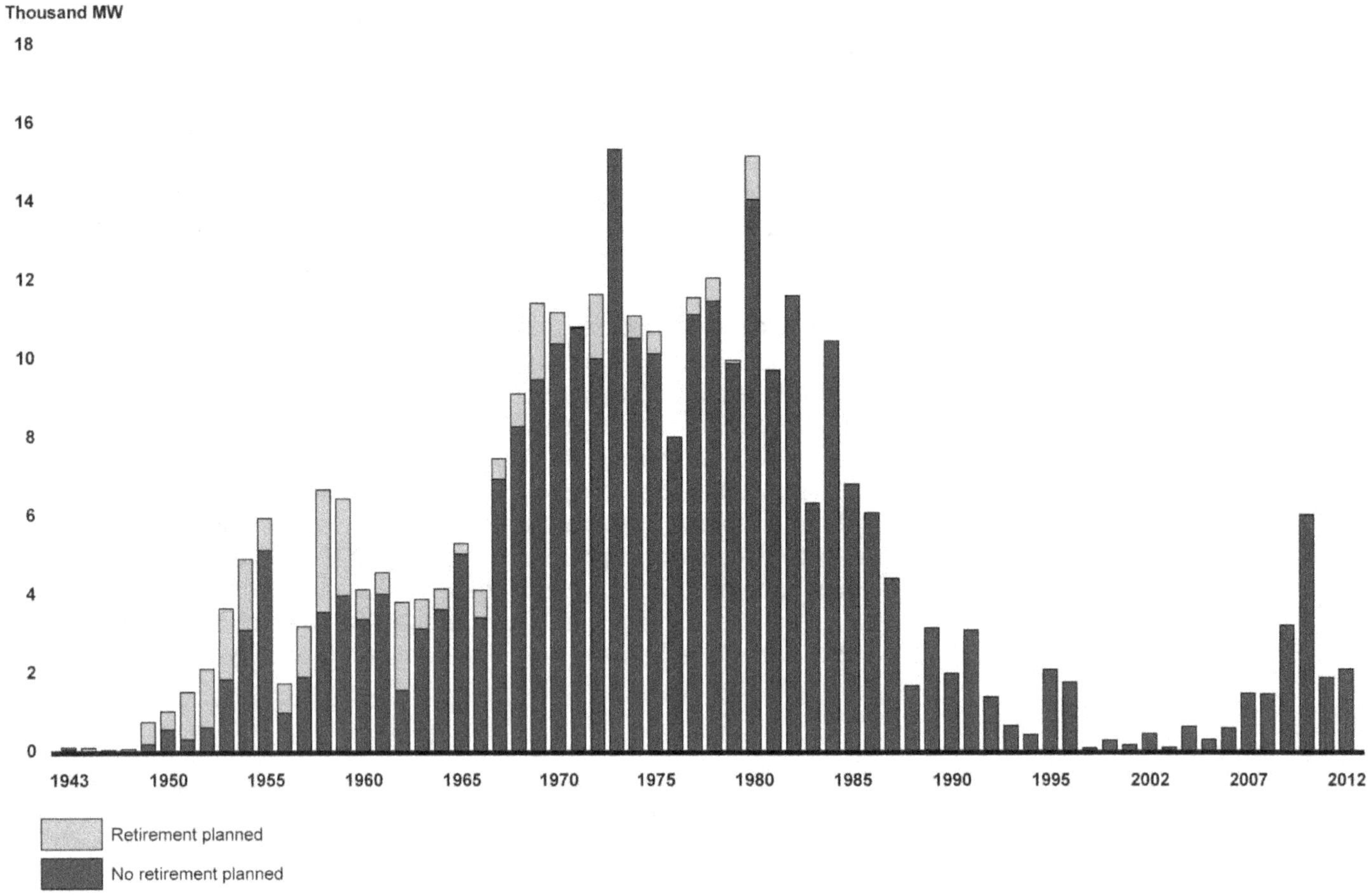

Source: GAO analysis of Ventyx data.

Note: Data are for coal-fueled electricity generating units greater than 25 MW net summer capacity, and planned retirements represent plans through 2020. Net summer capacity is a unit's capacity to generate electricity during the summer when electricity demand for many electricity systems and losses in efficiency are generally the highest. Net capacity excludes output used internally for plant operations.

- *Smaller.* The smaller a unit is, the more likely a power company is to be planning to retire it. (See fig. 5.) Size can be important when assessing the economics of additional investments needed to continue to operate coal-fueled units, as smaller units can be more expensive to retrofit, maintain, and operate on a per-MW basis. For example, some power companies may choose to install flue gas desulfurization units—known as scrubbers—to control SO_2 and other air emissions.[17] According to an EPA report, a typical 100 MW coal-fueled unit could incur capital costs 66 to 74 percent higher per MW to install a scrubber than a 700 MW unit.[18] In addition, smaller generating units are generally less fuel-efficient than larger units. Units that are planned for retirement average 175 MW of capacity compared with units that are not planned for retirement that average 351 MW of capacity. Figure 5 shows the number of coal-fueled units by capacity in MW.

[17]Scrubbers have been used commercially since the early 1970s and are the most common technology for reducing SO_2 emissions. They are capable of removing up to 99 percent of SO_2 emissions and work by injecting a sorbent—a material used to absorb molecules of a substance—into flue gas that reacts with pollutants to form a substance that is collected and removed. Scrubbers can also reduce emissions of mercury and other air pollutants. Another approach to reducing SO_2 emissions from coal-fueled electricity generating units is to switch from using coal with a higher sulfur content to coal with a lower sulfur content or to blend higher-sulfur coal with lower-sulfur coal. However, according to studies we reviewed, power companies may install scrubbers or other new control equipment to meet the requirements of new regulations.

[18]Specifically, according to the EPA report, capital costs could range from $385,000 to $470,000 per MW for a 700 MW unit, and from $641,000 to $817,000 per MW for a 100 MW unit. See: EPA, *Documentation for EPA Base Case v.4.10 Using the Integrated Planning Model*, EPA#430R10010 (Washington, D.C.: August 2010). (http://www.epa.gov/airmarkets/progsregs/epa-ipm/BaseCasev410.html)

Figure 5: Number of Coal-Fueled Electricity Generating Units by Capacity Category

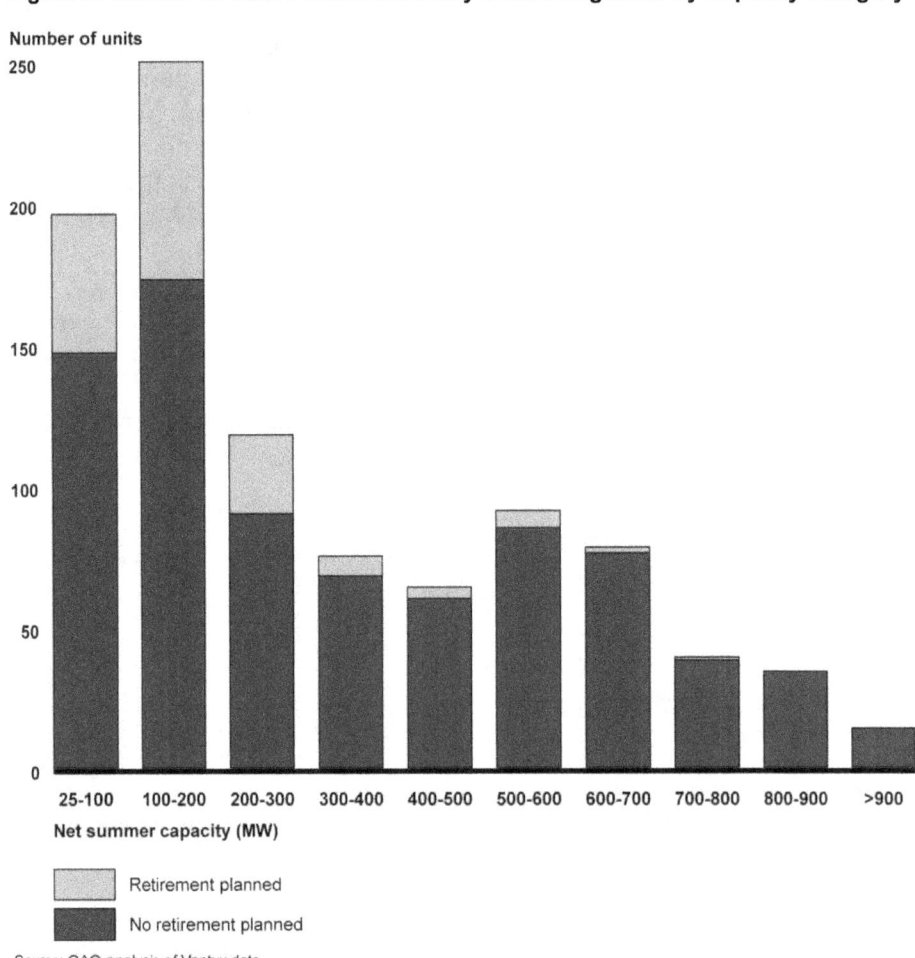

Source: GAO analysis of Ventyx data.

Note: Data are for coal-fueled electricity generating units greater than 25 MW net summer capacity, and planned retirements represent plans through 2020. Net summer capacity is a unit's capacity to generate electricity during the summer when electricity demand for many electricity systems and losses in efficiency are generally the highest. Net capacity excludes output used internally for plant operations.

- *More polluting.* Power companies' plans indicate they are more likely to retire units that emit SO_2 and NO_x at higher rates. Units in which pollution control equipment has been installed may require relatively minimal additional investments to meet new environmental regulatory requirements. For example, for units without controls to limit mercury emissions, power companies may have to install scrubbers or other controls, whereas units with such controls may already be able to meet new emissions limits. Fewer of the units that are planned for retirement have pollution control equipment and, therefore, units planned for retirement emit air pollutants such as SO_2 and NO_x at higher rates than the fleet overall. For example, 12 percent of the units planned for retirement have equipment installed to reduce SO_2 emissions, while almost 60 percent of units with no retirement plans have such equipment. As a result, units planned for retirement emitted an average of over twice as much SO_2 per unit of energy used in 2011 as units that are not planned for retirement—1.5 pounds of SO_2 for units planned for retirement compared with 0.6 pounds of SO_2 per million British thermal units (Btu) of energy used for units not planned for retirement (see fig. 6).[19] Similarly, units planned for retirement emitted on average about 60 percent more NO_x and 1 percent more CO_2 than units not planned for retirement.

---

[19]A Btu is a measure of energy that is the heat required to raise the temperature of 1 pound of water by 1 degree Fahrenheit.

Figure 6: Number of Coal-Fueled Electricity Generating Units by 2011 SO_2 Emissions Rate

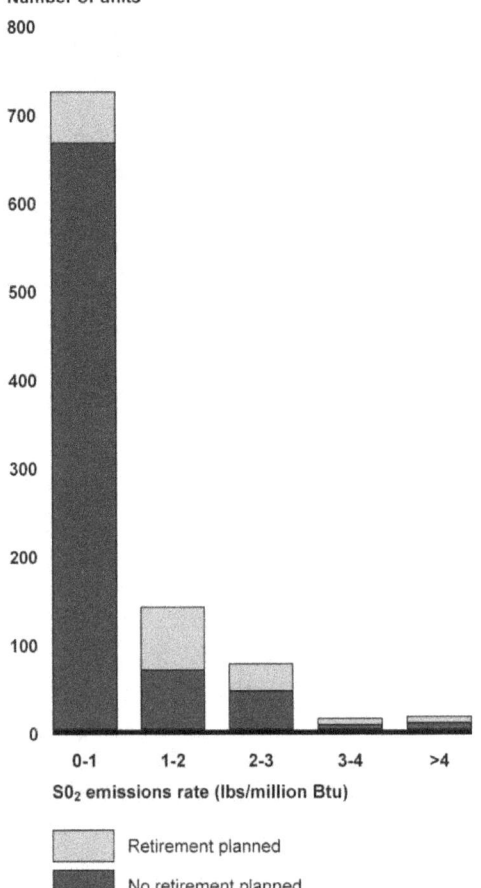

Number of units

Retirement planned

No retirement planned

Source: GAO analysis of Ventyx data.

Note: A British thermal unit (Btu) is a measure of energy and equals the heat required to raise the temperature of 1 pound of water by 1 degree Fahrenheit. Data are for coal-fueled electricity generating units greater than 25 MW net summer capacity, and planned retirements represent plans through 2020. Net summer capacity is a unit's capacity to produce electricity during the summer when electricity demand for many electricity systems and losses in efficiency are generally the highest. Net capacity excludes output used internally for plant operations.

Based on our statistical analysis of the characteristics associated with current retirement plans, we examined units owned by companies that have not announced any retirements to estimate the number of additional units and associated generating capacity that power companies may consider retiring.[20] Our analysis is subject to some uncertainty, and we therefore identified a range of units that power companies may consider retiring.[21] In addition to the 174 coal-fueled units with 30,447 MW of capacity (10 percent of total coal-fueled capacity in 2011) that are currently planned for retirement through 2020, our analysis predicts an additional 90 to138 coal-fueled units with 15,700 MW to 25,200 MW of capacity (5 to 8 percent of total) that companies may consider retiring. The capacity of units already planned for retirement (10 percent of total capacity), together with this additional capacity (5 to 8 percent of total capacity), suggests that 15 to 18 percent of total coal-fueled generating capacity could retire—an amount generally consistent with other forecasts we reviewed.

Many Units May Be Retrofitted with Pollution Control Equipment

As we reported in July 2012, power companies may retrofit many coal-fueled electricity generating units with new or upgraded pollution control equipment in response to new environmental regulatory requirements.[22] Though the requirements and deadlines these regulations may establish for generating units are somewhat uncertain at this time, EPA's analyses and two other studies we reviewed in our prior report suggest that one-third to three-quarters of all coal-fueled capacity could be retrofitted or upgraded with some combination of pollution control equipment, including scrubbers and other technologies to reduce SO_2, mercury, and other emissions. Once retrofitted with this pollution control equipment, the coal-fueled fleet would be capable of generating electricity and emitting much lower levels of pollution. For example, EPA projects that mercury emissions from coal-fueled electricity generating units will decrease by 75 percent as a result of its new regulatory requirements. Nevertheless, even the cleanest running coal-fueled unit may still be more polluting than generating units that use other fuel sources. For example, the 10 least-

---

[20]Specifically, we applied the results of our statistical analysis to units owned by power companies that have not made retirement announcements, assuming that they may still be assessing their options to identify units for retirement. We assume that companies that have announced some retirements have effectively announced all the units they plan to retire. See appendix I for additional information on our analysis.

[21]Specifically, we identified the 95 percent confidence interval in that our analysis suggests that there is a 95 percent probability that the actual number of units that may retire is within this range.

[22]See GAO-12-635.

emitting coal-fueled units emitted over 10 times as much SO_2 per unit of energy input than the average combined cycle natural gas unit in 2011— an average of 0.007 pounds of SO_2 per million Btu compared with an average of 0.0006 for combined cycle units.[23] Electricity generating units that rely on solar and wind sources produce no such emissions.

Some New Generating Units May Be Built and Would Be Larger, Cleaner, and More Efficient Than the Fleet Overall

Available information suggests that industry intends to build some new coal-fueled electricity generating units. According to Ventyx data, power companies have plans to build 42 new coal-fueled electricity generating units with 21,634 MW of capacity in various stages of planning or development (see fig. 7). However, as we have previously reported, developers generally have more planned projects than they complete.[24]

---

[23]Similarly, the 10 cleanest coal-fueled units emitted 24 percent more NO_x and 79 percent more CO_2 per million Btu than the average combined cycle natural gas unit. Many natural gas-fueled units built recently have used highly efficient combined-cycle technologies, which rely on large gas turbines, also called combustion turbines, together with a steam generator and a steam turbine to convert waste heat in the exhaust stream to electricity.

[24]GAO, *Restructured Electricity Markets: Three States' Experiences in Adding Generating Capacity,* GAO-02-427 (Washington, D.C.: May 24, 2002).

Figure 7: Capacity of Planned Coal-Fueled Electricity Generating Units by Status, as of July 26, 2012

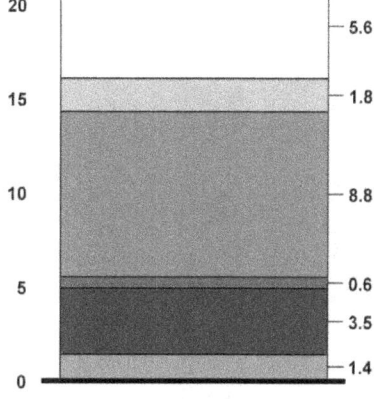

Thousand MW

- 5.6
- 1.8
- 8.8
- 0.6
- 3.5
- 1.4

☐ Proposed (new generating unit planned)

▨ Application pending (permits and regulatory approvals pending)

▨ Permitted (two or more permits have been approved or contracts received)

▨ Site prep (site is being prepared for construction)

▨ Under construction (unit under construction)

▨ Testing (operating under test conditions)

Source: GAO analysis of Ventyx data.

Note: Data refer to net summer capacity—the generating unit's capacity to generate electricity during the summer when electricity demand for many electricity systems and losses in efficiency are generally the highest. Net capacity excludes output used internally for plant operations.

The total capacity of coal-fueled electricity generating units in the United States may decline in the future as less capacity is expected to be built than is expected to retire. As discussed, 49,000 to 76,476 MW of coal-fueled capacity is projected to retire by 2035 according to EIA and IHS Global Insight, respectively, and they project that 11,000 MW and 22,134 MW of new coal-fueled capacity will be added by 2035, respectively. EIA officials told us that new coal-fueled capacity in their projections is primarily expected in the next few years and represents units that are already planned or under construction. As less capacity is expected to be built than is expected to retire, total coal-fueled capacity is expected to decline in the future, as shown in figure 8. Coal's share of total electricity generating capacity was about 30 percent in 2011. In EIA's reference scenario, coal's share of capacity declines to 25 percent in 2035 as retiring coal-fueled units are not fully replaced, and as 176,100 MW of other generating capacity is added in the future.

Figure 8: Actual 2011 and Projected 2035 Coal-Fueled Electricity Generating Capacity

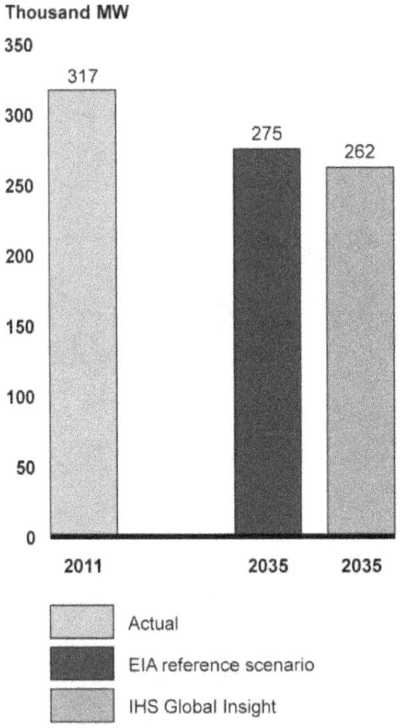

Thousand MW

Legend:
- Actual
- EIA reference scenario
- IHS Global Insight

Sources: GAO analysis of EIA reference scenario and IHS Global Insight data.

Any coal-fueled units that are built in the future are likely to be larger, less polluting, and more fuel-efficient than the average of the coal-fueled fleet overall. Units that power companies are currently planning to build average 515 MW of net summer capacity, and the operating fleet averages 319 MW.[25] Additionally, new units must install technologies to control emissions, and so are likely to emit lower levels of pollutants and thus be cleaner than the fleet overall. For example, generating units built after August 7, 1977, have had to obtain preconstruction permits that establish air emissions limits and require the use of certain emissions

---

[25]As discussed, a number of generating units are expected to be retired in the future, and these tend to be smaller units. Of the operating fleet, the average size of units planned for retirement is 175 MW, and the average size of units without retirement plans is 351 MW.

control technologies such as scrubbers to reduce emissions of SO_2.[26] In addition, some stakeholders we interviewed said that new coal-fueled units were likely to incorporate designs that are able to convert fuel to electricity more efficiently.

Coal Likely to Remain a Key Fuel Source, but Future Use May Be Affected by Fuel Prices, Environmental Regulations, and Other Factors

Coal is likely to continue to be a key fuel source for electricity generation in the United States, but its share as a source of electricity is expected to decline, and the future use of coal to generate electricity in the United States may be affected by several key factors that include the price of natural gas and other competing fuels, environmental regulations, and the demand for electricity, among others. In addition, several stakeholders we interviewed said that coal may increasingly be exported for use in other nations, though the extent of future exports is uncertain.

Coal Likely to Continue to Be a Key Source of Electricity in the Future, though Its Share Is Generally Expected to Decline in the United States

According to stakeholders we interviewed and projections by EIA, IEA, and IHS Global Insight, coal is likely to continue to be a key fuel source for U.S. electricity generation, but its share as a source of electricity is generally expected to decline in the future. Some stakeholders told us that, in the future, electricity generation from coal is likely to be displaced by generation from other fuel sources, particularly natural gas, but they still expect coal's contribution to electricity generation to be significant. Furthermore, in its reference scenario, EIA estimates that coal will represent 38 percent of U.S. electricity generation in 2035 under current policies—down from 42 percent in 2011.[27] As shown in figure 9, the total amount of electricity generated using coal is expected to remain relatively constant over this same period under EIA's reference scenario, growing by 0.1 percent annually. However, the amount of electricity generated using some other fuel sources, for example, natural gas and renewables, will increase at higher annual rates—1.4 percent and 2.3 percent respectively—diminishing coal's total share of electricity generation.

---

[26]The New Source Review provisions of the Clean Air Act establish this permitting process. See GAO-12-545R.

[27]As mentioned, coal's share of total electricity generating capacity was about 30 percent in 2011 and is expected to decline, according to EIA's reference scenario, to 25 percent in 2035 as retiring coal-fueled generating units are replaced with few new coal-fueled units.

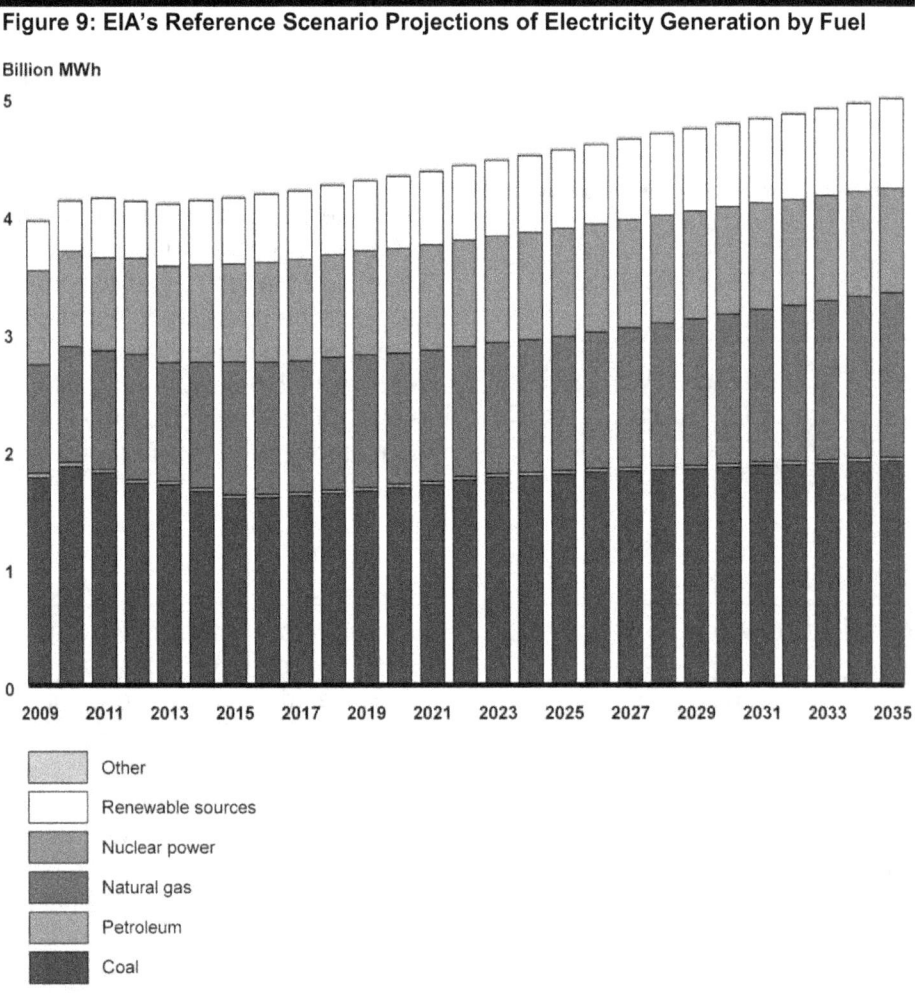

Figure 9: EIA's Reference Scenario Projections of Electricity Generation by Fuel

Billion MWh

Legend:
- Other
- Renewable sources
- Nuclear power
- Natural gas
- Petroleum
- Coal

Source: EIA.

Note: Data from 2009 and 2010 represent historical data. Data from 2011 and after represent EIA estimates. Renewable sources include conventional hydroelectric, geothermal, wood, wood waste, certain municipal wastes, landfill gas, other biomass, solar, and wind power. Other includes pumped storage, certain municipal wastes, refinery gas, still gas, batteries, chemicals, hydrogen, pitch, purchased steam, sulfur, and miscellaneous technologies.

IEA, in its current policies scenario, which considers policies enacted by mid-2011, projects that coal's share of electricity generation will increase slightly to 43 percent in 2035. IHS Global Insight, which assumes some changes in current U.S. policies in its analysis, projects that coal's share of electricity generation will decline to 26 percent in 2035.

Power companies are expected to retire a significant number of coal-fueled generating units in the future, but these retirements may have more of an impact on coal-fueled capacity than on electricity generation from coal. As discussed, our statistical analysis suggests that 15 to 18 percent of total coal-fueled generating capacity could retire. Units that may retire did not run as intensively as coal-fueled units overall, and we estimate they may account for 13 to 16 percent of the annual average electricity generation from coal.[28] (See app. I for additional information on our statistical analysis.)

The changes in coal use may also result in shifts between major coal-producing areas in the United States. As shown in figure 10, EIA identifies three broad coal-producing regions in the United States: Appalachia, the Interior region spanning coal-producing areas in central states to Texas, and the Western region covering coal-producing areas in western states and Alaska. In EIA's reference scenario, coal production from Appalachia declines, and production from the Western and Interior regions increases through 2035.[29] According to EIA, in 2010, 31 percent of coal was produced in Appalachia, 14 percent was produced in the Interior region of the United States, and 55 percent was produced in the West. As shown in figure 10, EIA's reference scenario projects that these production figures will change by 2035, with 24 percent of coal produced in Appalachia, 16 percent produced in the Interior region, and 60 percent produced in the Western region.

---

[28]Generation is based on annual average generation 2007-2011.

[29]Of the three estimates we reviewed, EIA was the only one that provided information on the location of coal production in the United States.

Figure 10: Actual 2010 and EIA's Reference Scenario Projected 2035 Coal Production by Region

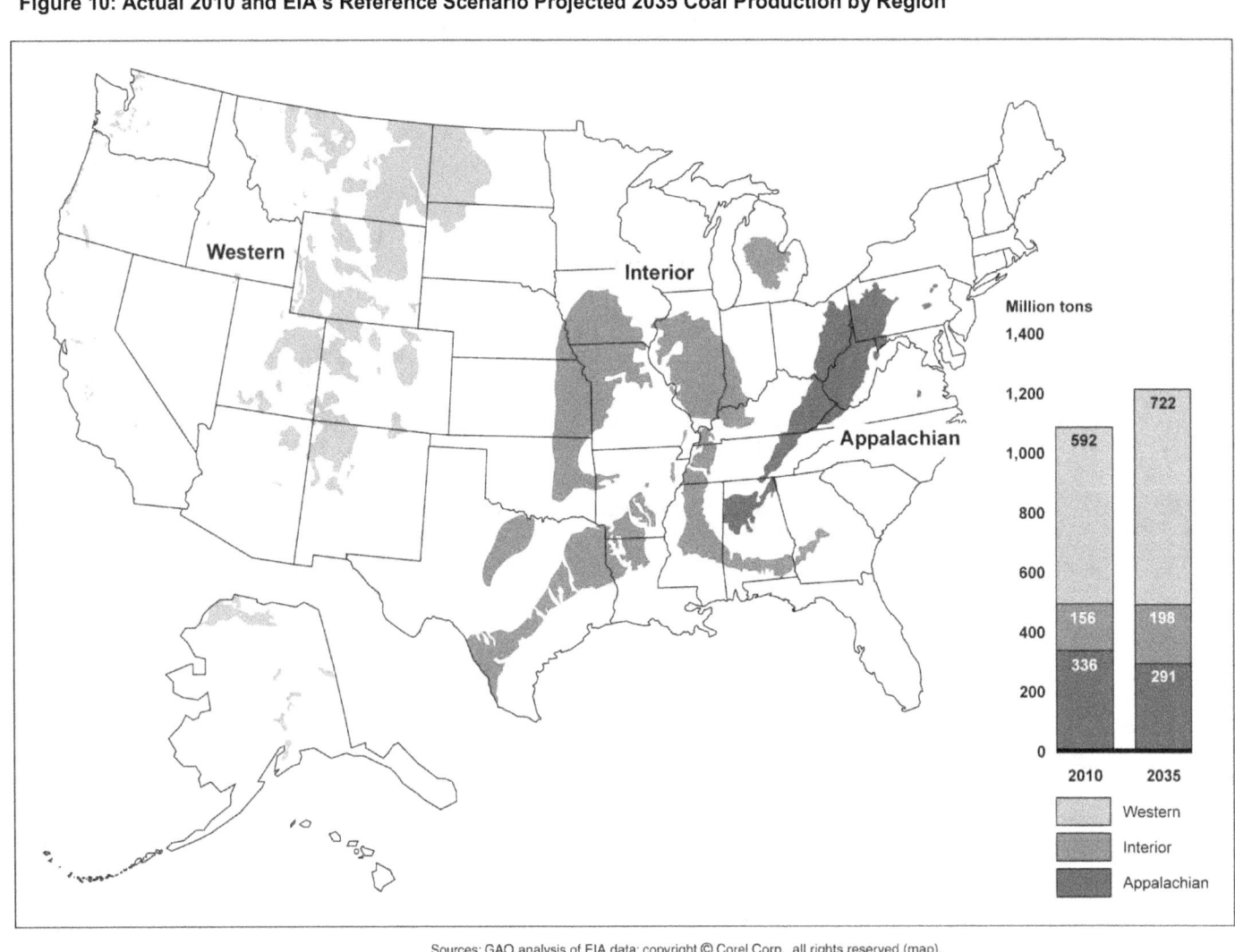

Within Appalachia, EIA expects declines to come from the central region, which includes southern West Virginia, Virginia, eastern Kentucky, and northern Tennessee. This expected shift in coal production from the eastern United States to the West represents an industry trend ongoing since the early 1990s that is influenced by each region's unique set of complex geological, mining, and transportation characteristics. For example, some stakeholders told us that demand for western coal has increased primarily because it is low in sulfur content, and the region's coal reserves can be mined relatively inexpensively compared with Appalachian and Interior coal reserves, which are often more deeply underground and costlier to access. Available information suggests that these benefits have made western coal economically competitive with coal from the Appalachian and Interior regions, despite western coal's lower heating value and higher cost to transport to some coal-fueled generating units.

Future Fuel Prices, Environmental Regulations, and Other Factors May Affect the Use of Coal in the United States

According to available information, the future use of coal to generate electricity in the United States may be determined by several key factors.

Prices for Natural Gas and Other Competing Fuels

Available information suggests that the price of coal compared to the prices of other fuel sources could influence how economically attractive it is to use coal to generate electricity. This can affect how often existing coal-fueled generating units are used, how many units are retired, and how many new units are built. In general, the decision about whether to operate a given generating unit is based on the costs of operating that unit. Operating costs are driven, in part, by the cost of fuel sources and how efficiently fuels are converted into electricity. In general, new natural gas-fueled generating units are able to convert fuel into electricity more efficiently than existing coal-fueled generating units, meaning they can convert a unit of fuel energy into more electricity than less-efficient coal-fueled units. Some natural gas-fueled units constructed in the last decade can require less than 7,000,000 Btus of natural gas to generate one MWh of electricity. In contrast, existing coal-fueled generating units require around 10,000,000 Btus of coal to generate one MWh of electricity, and 187 coal-fueled units require over 12,000,000 Btus per MWh. Newer designs of coal-fueled units exist that can operate at higher efficiencies, but few have been built in the United States.

Generally, generating units with the lowest costs operate more often than units with higher costs. If the price of natural gas falls relative to coal in a particular region, depending on each unit's efficiency, it may result in the operating costs of some natural gas units dropping below the operating costs of some coal units and, thus, natural gas units being operated more often—and coal units less often—than before. We previously reported that, in some areas of the country, it has become less economically attractive to use coal to generate electricity, as the regional prices of coal have increased, the prices of natural gas have fallen, and the availability of natural gas has increased.[30] Multiple stakeholders told us that if natural gas prices remain low relative to coal prices, this trend could continue. As shown in figure 11, prices of coal and natural gas have varied historically, and EIA and IHS Global Insight project a range of potential future prices in their forecast scenarios.

Figure 11: Actual and Projected Coal and Natural Gas Prices

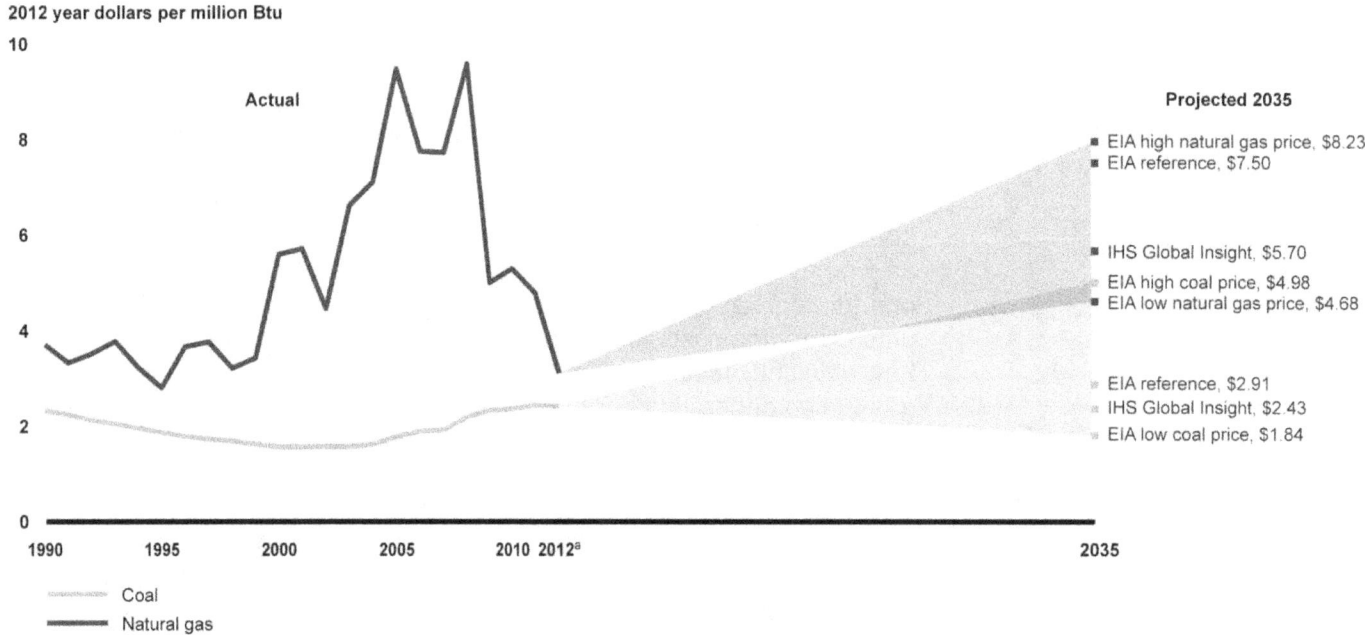

Sources: GAO analysis of EIA and IHS Global Insight data.

Note: Shading is not indicative of projected prices between 2012 and 2035. As described, both fuel prices and efficiency will affect a generating unit's operating costs and, thus, how often one unit operates compared with another. Many newer natural gas-fueled units are more efficient than coal-fueled units, and when the price of natural gas drops to a certain level relative to coal prices, it becomes less costly to operate these efficient natural gas units than coal units. Data for the high price scenarios were obtained from EIA's "Low Estimated Ultimate Recovery" and "High Coal Cost" scenarios. These scenarios had the highest prices for natural gas and coal delivered to the electricity sector after the two greenhouse gas scenarios discussed later. Data for the low price scenarios were obtained from EIA's "High Total Recoverable Resources" and "Low Coal Cost" scenarios. Data were converted to 2012 year dollars using the gross domestic product deflator based on the calendar year.

[a]Average through June 2012.

---

[30]GAO-12-635.

EIA's forecast suggests that future fuel prices may influence the extent to which coal is used to generate electricity. As shown in Table 1, four EIA scenarios project that the share of total electricity generated using coal could vary from 30 to 42 percent in 2035 based on fuel prices. See appendix II for further information about these scenarios.

Table 1: Results of EIA Alternate Fuel Price Scenarios

Scenario	Electricity generated using coal (million MWh)	Average annual percent change in electricity generated using coal, 2010–2035 (percentage)	Share of total electricity generated using coal
Actual, 2010	1,851	Not applicable	45
Actual, 2011	1,734	Not applicable	42
Reference Scenario, 2035	1,897	0.1 %	38
Scenario with High Coal Prices, 2035	1,577	-0.6%	32
Scenario with Low Coal Prices, 2035	2,107	0.5%	42
Scenario with High Natural Gas Prices, 2035	1,977	0.3%	40
Scenario with Low Natural Gas Prices, 2035	1,542	-0.7%	30

Source: EIA, *2012 Annual Energy Outlook*. Actual 2011 data are from EIA's August 2012 *Electric Power Monthly*.

Note: Data for the high price scenarios were obtained from EIA's "Low Estimated Ultimate Recovery" and "High Coal Cost" scenarios. These scenarios had the highest prices for natural gas and coal delivered to the electricity sector after the two greenhouse gas scenarios discussed later in this report. Data for the low price scenarios were obtained from EIA's "High Total Recoverable Resources" and "Low Coal Cost" scenarios.

Current and Future Environmental Regulations and Renewable Energy Policies

Available information indicates that existing and potential future regulations may make it more expensive to generate electricity using coal, thus affecting coal's future use. Some stakeholders we spoke with indicated that current and proposed regulations addressing air pollution, water pollution, and wastes from using coal could adversely affect coal in the future by making it more expensive to generate electricity using coal.[31] Some stakeholders also said that coal would be adversely affected by efforts to reduce greenhouse gas emissions from the electricity industry, such as by a regulation EPA proposed in 2012—new source performance standards for greenhouse gas emissions, which would establish limits on the amount of CO_2 that new generating units can emit—as well as the potential for additional federal efforts to regulate greenhouse gases.[32] Available information also indicates that the extent to which these regulations adversely affect coal will depend, in part, on the future economics and development of technologies to address CO_2 emissions from coal-fueled units, such as technologies to capture and store CO_2. The development of effective and commercially viable CO_2 controls for coal-fueled electricity generating units has received significant attention, but some of these technologies are still in the research and development phase, and most are not yet commercially viable.[33]

Though it is unclear whether there will be additional future efforts to reduce CO_2 emissions, EIA assessed two scenarios examining

---

[31]Stakeholders provided various examples of current environmental and other regulations with the potential to affect the coal industry in the future, including the National Emissions Standards for Hazardous Air Pollutants from Coal and Oil Fired Electric Utility Steam Generating Units, also known as the Mercury and Air Toxics Standards, which establishes emissions limitations on mercury and other toxic pollutants. Stakeholders also discussed the proposed Cooling Water Intake Structures at Existing Facilities and Phase I Facilities regulation, which would establish requirements for water withdrawn and used for cooling purposes that reflect the best technology available to minimize adverse environmental impact and the proposed Disposal of Coal Combustion Residuals from Electric Utilities regulation, which would govern the disposal of coal combustion residuals, such as coal ash, in landfills or surface impoundments.

[32]EPA has taken actions to regulate greenhouse gas emissions from electricity generating units. Specifically, since 2011, EPA regulations have required permits for greenhouse gas emissions from certain new and existing units. In addition, under the terms of a settlement agreement, EPA was required to propose and finalize new source performance standards for greenhouse gas emissions from new and existing electricity generating units. In April 2012, EPA proposed standards for certain new electricity generating units and stated that it would propose standards for existing units at an appropriate time.

[33]In 2008, we reported on the key economic, legal, regulatory, and technological barriers impeding commercial-scale deployment of carbon capture and storage technology and actions certain federal agencies are taking to overcome these barriers. GAO, *Climate Change: Federal Actions Will Greatly Affect the Viability of Carbon Capture and Storage As a Key Mitigation Option*, GAO-08-1080 (Washington D.C.: Sept. 30, 2008). In 2010, we reported on the maturity, commercial viability, and implications of technologies to reduce carbon dioxide from coal power plants. GAO, *Coal Power Plants: Opportunities Exist for DOE to Provide Better Information on the Maturity of Key Technologies to Reduce Carbon Dioxide Emissions*, GAO-10-675 (Washington, D.C.: June 16, 2010).

hypothetical future policies. One scenario has a hypothetical price on CO_2 emissions starting at $15 per ton in 2013 that increases 5 percent per year and that would reduce CO_2 emissions from the electric power sector by 46 percent below 2010 levels in 2035. Another scenario has an initial price on CO_2 emissions of $25 per ton that increases 5 percent per year and would reduce CO_2 emissions from the electric power sector by 76 percent below 2010 levels in 2035. EIA stated that these scenarios are consistent with previously proposed legislation. EIA projected that, under these two scenarios, coal's share of total electricity generation could fall to 16 and 4 percent in 2035, respectively. IEA also developed scenarios that took potential additional efforts to address climate change into account—one scenario that considers the United States' international commitments, even if nonbinding, and a second scenario that considers an approach to energy use in the United States consistent with more significant future steps to address climate change. Under these scenarios, IEA estimates that 35 and 17 percent of U.S. electricity would be generated by coal, respectively.[34]

In addition, many states have policies in place requiring power companies to increasingly rely on renewable sources for electricity, and similar proposals have been introduced at the federal level. We have previously reported that 30 states have laws or regulations requiring power companies to increasingly rely on renewable sources for electricity.[35] At the federal level, there have been several efforts to establish a national renewable portfolio standard. For example, a Senate bill introduced in March 2012 would require power companies to obtain a certain amount of the electricity they sell from clean energy sources with zero or low carbon generation, such as renewable energy, natural gas, and nuclear power. An EIA analysis of this bill suggests that such a policy could increase

[34]Under the first scenario, IEA estimates a reduction of U.S. CO_2 emissions from energy of 12 percent by 2035 compared with emissions in 2009. The second scenario represents a reduction of U.S. CO_2 emissions from energy of more than half by 2035 compared with emissions in 2009.

[35]These states are: Arizona, California, Colorado, Connecticut, Delaware, Hawaii, Illinois, Iowa, Kansas, Maine, Maryland, Massachusetts, Michigan, Minnesota, Missouri, Montana, Nevada, New Hampshire, New Jersey, New Mexico, New York, North Carolina, Ohio, Oregon, Pennsylvania, Rhode Island, Texas, Washington, West Virginia, and Wisconsin.

demand for renewables and other sources favored by these rules and diminish demand for electricity from coal.[36]

Demand for Electricity

Available information suggests that future electricity demand levels may affect the use of coal to generate electricity. In general, if more electricity is needed in the future, it becomes more likely that existing generating units that are not already running at capacity—including coal units—will be operated more often. Lower demand makes it more likely that the opposite will happen. Demand levels also influence decisions about whether to build new generating units, including coal-fueled units, with high demand scenarios increasing the likelihood that new units will be operated frequently enough to recover the capital costs associated with their construction. According to EIA, electricity demand is affected by factors including population growth and economic growth. In general, electricity use increases as the economy grows, but improvements in energy efficiency can offset some or all of the increases in electricity use due to economic growth.

Electricity demand varies across EIA's scenarios. In its high economic growth scenario, in which future electricity demand is the highest, total demand for electricity generated using coal grows at an annual rate of 0.3 percent from 2010 to 2035 to 2,004 million MWh in 2035, higher than the 1,897 million MWh estimate from EIA's reference scenario. However, in this scenario, EIA does not predict a significant change in coal's share of total electricity generation—37 percent compared with 38 percent in the reference scenario. EIA's scenario that assumes high use of energy efficient technologies in the residential and commercial sectors estimates the lowest future demand for electricity. In this scenario, electricity generated from coal declines at an annual rate of 1.1 percent to 1,414 million MWh in 2035, and electricity generated from coal constitutes 34 percent of total electricity generation that year.

Other Factors

A variety of other factors may also affect the future of coal in the United States. For example, some stakeholders told us that local opposition to coal by environmental activists and other concerned citizens has and will continue to adversely impact its future use. According to some other stakeholders, coal's future will also be influenced by the extent to which technologies to reduce coal's environmental impact become commercially

---

[36]EIA, *Analysis of the Clean Energy Standard Act of 2012* (Washington, D.C.: May 2012).

available and economic. For example, technology to capture and store carbon underground has the potential to address the adverse greenhouse gas impacts of electricity generation using coal, which could increase demand for coal.

Extent of Future Coal Exports Is Uncertain

Several stakeholders we interviewed said that U.S. coal may increasingly be exported for use in the electricity sector and for other purposes, but the extent of future U.S. coal exports is uncertain. According to IEA, world demand for coal grew rapidly from 2000 to 2010, and coal accounted for nearly half the increase in global energy use over that period. Additionally, according to IEA, world coal demand is strongly correlated with global economic activity and will also be affected by world energy and environmental regulations and the development and use of technologies to reduce coal's environmental impact.

Available information suggests that the future level of U.S. coal exports will also depend on how competitive U.S. coal prices are internationally and the extent to which the quality of coal available from the United States is in demand. For example, metallurgical coal—coal used for such activities as steel production—has historically constituted a significant share of U.S. coal exports. Factors affecting the ability of other coal-exporting countries to economically and reliably supply coal to international customers include local freight rates, limits on the amount of exports, and extreme weather events—all of which can influence the relative price of U.S. coal and, thus, the amount of U.S. coal exported.

According to EIA's reference scenario, coal exports are estimated to grow 1.8 percent annually from 2010 to 2035, reaching 129 million tons of coal (11 percent of total U.S. production) in 2035. On the other hand, IHS Global Insight projects that exports will fall 1 percent annually from 2010 to 2035, with 62.8 million tons of coal being exported in 2035 (7 percent of total U.S. production).

Agency Comments and Our Evaluation

We met with EIA officials to discuss an early draft of this report and incorporated technical suggestions where appropriate. We also provided a draft of this report to EIA and EPA for formal comment. EIA and EPA did not provide written comments for inclusion in this report. EPA's Office of Air and Radiation did provide technical comments and stated that the report contained a very good description of many of the changes going on in coal and electricity markets that are affecting the use of coal to generate electricity. In its technical comments, EPA suggested that the draft's emphasis on environmental regulations, particularly on the Highlights page, was misleading and not consistent with the rest of the report, which has a fuller discussion of many factors affecting the future use of coal. EPA stated that market changes, which we discuss in the report, would have significant impacts even in the absence of EPA's regulations. We do not agree that the report was misleading, but given that the Highlights page may be read without the benefit of the fuller discussion found in the report, we moved language from the body of the report to the Highlights page about other factors affecting the use of coal. EPA provided other technical comments, which we incorporated where appropriate.

As agreed with your office, unless you publicly announce the contents of this report earlier, we plan no further distribution until 30 days from the report date. At that time, we will send copies to the appropriate congressional committees, the Administrators of the EIA and EPA, and other interested parties. In addition, the report will be available at no charge on the GAO website at http://www.gao.gov.

If you or your staff members have any questions about this report, please contact me at (202) 512-3841 or ruscof@gao.gov. Contact points for our Offices of Congressional Relations and Public Affairs may be found on the last page of this report. GAO staff who made key contributions to this report are listed in appendix III.

Sincerely yours,

Frank Rusco
Director, Natural Resources and Environment

Appendix I: Analysis of Characteristics of Coal-Fueled Generating Units That Power Companies Plan to Retire

This appendix describes our statistical analysis of characteristics of coal-fueled electricity generating units, such as age and size, that are likely to affect power companies' plans to retire certain units. We use this analysis to estimate the number and generating capacity of other coal-fueled units that power companies are likely to consider retiring.

Methodology

To test the hypothesis that power companies are likely to retire older, smaller, and more polluting coal units by 2020, we used logistic regression analysis. We analyzed industry data on all coal-fueled units owned by power companies that have already announced plans to retire one or more of these units. Using unit- and company-level data, primarily from company-reported databases, we developed a model depicting the relationship between companies' announced plans to retire a unit and that unit's characteristics—age, size, emissions rates of sulfur dioxide (SO_2) and nitrogen oxides (NO_x), and the regulatory status of the power company that owns the unit, specifically whether the company is traditionally regulated or operates in a restructured market. To estimate the number and generating capacity of additional units likely to be retired, we applied our model to a dataset consisting of coal-fueled units owned by power companies that have not announced any retirements.

Model of Plans to Retire Coal-Fueled Generating Units

In developing our model of power companies' plans to retire coal-fueled units, we relied on economic theory, as well as discussions with stakeholders and our review of studies. Stakeholders included representatives from power companies, a coal company, industry associations, and nongovernmental organizations, and officials from federal and state agencies. Stakeholders and studies mentioned the following characteristics as likely unit-level determinants of power companies' plans to retire a coal-fueled unit or keep it in operation:

- age;
- generating capacity;
- fuel efficiency (i.e., how efficiently a unit converts fuel to electricity)
- operating cost and profitability;
- pollution emission rates and whether a unit already has various types of emissions control equipment; and
- regulatory status.

As a general matter, the larger, newer, more efficient, and less polluting a generating unit is, the more likely it is that a power company may to want to keep it in service and invest in retrofits that may be needed for it to comply with environmental laws or regulations. For example, if a large, new generating unit that a power company uses to meet a significant portion of customer demand is not in compliance with environmental regulations, retiring it would likely require replacing it with another unit of similar size. Doing so may be very costly, and retrofitting it with the requisite pollution control equipment may be a more economical choice.

It is also reasonable to expect regulatory status to have some impact on power companies' retirement plans because such plans could involve significant investments. For companies that are traditionally regulated, state public utility commissions review power companies' plans for major investments in pollution control equipment in the case of a retrofit, or in replacement power generation capacity if it is needed after a unit is retired. Decisions by power companies in restructured markets are not subject to the same state public utility oversight. Furthermore, once state public utility commissions approve a traditionally regulated company's plan to invest in major retrofits or replacement units, they allow it to charge rates to recover its investment costs. Companies operating in restructured markets have no such cost-recovery provisions, so their investments in retrofits or replacement units may be riskier.

Our model does not include all the characteristics that stakeholders and studies identified as possible characteristics that power companies consider in deciding which coal-fueled units to retire. First, economic theory and our analysis of data on coal-fueled units indicate that there are interrelationships among some of these characteristics; for example, newer, larger electric generating units tend to be more fuel efficient, and this fuel efficiency contributes to lower operating costs. Hence, including all characteristics would be redundant and weaken the statistical results. Below, we discuss some specifications of the model with alternative sets of variables. Second, there are likely other characteristics that may influence power companies' plans to retire generating units that we were unable to include in our statistical analysis. We discuss limitations of our model below.

Data Used

We used U.S. electricity data at the level of individual coal-fueled generating units that we obtained under contract from Ventyx,[1] a company that maintains a proprietary database containing consolidated energy and emissions data from the Energy Information Administration (EIA),[2] the Environmental Protection Agency (EPA), and other sources. In particular, we relied primarily on two datasets from Ventyx, "Generating Unit Capacity," and "Unit Generation & Emissions – Annual," from which we assembled our dataset of characteristics of operating coal-fueled generating units. These characteristics included some of the following:

- generating unit identification data, including name of power company owning the unit, location, and unit identification numbers;
- proposed retirement year;
- age;
- size, measured in megawatts (MW) of generating capacity;
- fuel efficiency;
- emissions rates of SO_2, NO_x, and carbon dioxide;
- types of installed control equipment or whether owners plan to install control equipment in the future;
- various cost measures, including generating unit marginal cost; and
- regulatory status: equals 1 if the power company that owns the unit was traditionally regulated or 0 if the company was operating in a restructured market.[3]

We also used regional day-ahead market prices from the IntercontinentalExchange (ICE) company, and spot market prices from the Federal Energy Regulatory Commission (FERC) to calculate an average wholesale market price for the regional markets associated with

---

[1]Ventyx is an operating unit of the ABB Company.

[2]EIA is a statistical agency within the Department of Energy that collects, analyzes, and disseminates independent information on energy issues.

[3]In its datasets, Ventyx uses the term "unregulated" instead of the term we use, "restructured." We use the term "restructured" because all generating units are subject to some type of government regulation, even if they are not traditionally regulated. A limited number of coal-fueled units in our sample are jointly owned by two or more companies with at least one owner being traditionally regulated and at least one other operating in a restructured market. In these cases, we designated the unit owner's regulatory status according to the majority owner. In the case of two power companies owning six units, ownership was 50/50 between companies that were traditionally regulated and those operating in a restructured market. In this case, we decided to use the designation traditionally regulated for the power company's regulatory status.

each unit in our dataset.[4] For each market region, we calculated a simple average of daily prices for the year 2011 from daily ICE price data. For some of the regions, however, there were no price data available from ICE, so we used the 2011 average spot market price from FERC.[5]

While our model does not include all the aforementioned characteristics, we used most of these characteristics in alternative specifications of the model and discuss two of these specifications below.

Our complete dataset includes 959 coal-fueled units. This dataset includes only units that have a net summer generating capacity greater than 25 MW, making them subject to EPA emissions monitoring and reporting requirements. We excluded units that have not reported any electricity generation or SO_2 or NO_x emissions over the past 5 years.[6] Of the total 959 units, 482 units belong to power companies that have announced plans for retiring at least one coal-fueled unit.

Results

We used logistic regression (logit) analysis to analyze the characteristics that are affecting power companies' retirement plans of coal-fueled electricity generation units. Regression analysis in general estimates the effect of a change in an independent variable on the outcome (dependent) variable, while holding other variables constant. Logit is a type of regression analysis for situations in which the dependent variable is a categorical variable—one that can take on a limited number of values—instead of a continuous, quantitative variable. In this case, the categorical variable is binary, which means that the choice is between only two outcomes.

---

[4]ICE is a company that offers commodity trading services in energy, agriculture, and other sectors. ICE day-ahead electricity prices are publicly available on ICE's website. While the Ventyx datasets to which we had access do not include regional wholesale electricity price data, they do indicate the ICE electricity market regions (referred as "hubs") relevant to each of the units in the datasets that we used. The prices that FERC reports are based on data from *Platt's,* a leading provider of commodity markets data.

[5]We compared the average 2011 prices from the ICE data source and from FERC for the regions that both sources reported, and they were extremely close.

[6]Net summer capacity is a unit's capacity to produce electricity during the summer when electricity demand for many electricity systems and losses in efficiency are generally the highest. Net capacity excludes output used internally for plant operations.

We estimated the logit regression equation for the subgroup of 482 coal-fueled generating units belonging to power companies that have announced plans to retire at least one coal-fueled unit. The dependent variable in our model is whether to retire or not retire a coal unit, and the independent variables are the (1) age of unit; (2) net summer capacity as a measure of unit size; (3) unit's SO_2 emissions rate in pounds (lb) of SO_2 emissions per unit of heat input from the fuel used in the unit's electricity generation, measured using millions of British thermal units (Btu); (4) unit's NO_x emissions rate in lb/million Btu; (5) whether the power company that owns the unit is traditionally regulated or operates in a restructured market.

Table 2 shows our resulting estimated equation and relevant statistics.

Table 2: Regression Results of Characteristics Associated with Planned Coal-Fueled Unit Retirements–Dependent Variable Is the Probability of a Unit Retirement Announcement

Variable	Coefficient	Standard error	Significance[a]
Net summer capacity	-0.00198	0.0010	5.9%
Unit age	0.06647	0.0201	0.1%
SO_2 emissions rate	0.68053	0.1550	0.0%
NO_x emissions rate	3.64861	0.9168	0.0%
Regulatory status	-1.56564	0.3331	0.0%
Constant	-3.84944	1.3276	0.4%

Source: GAO analysis of Ventyx data

Note: Number of observations equals 482.

[a]Significance is measured by the probability that we would find these results if the true value of the coefficient were zero. A lower significance level means that it is less likely that the observed pattern in the data is the product of pure chance. As with any regression, the statistical significance depends on the model being correctly specified.

These results generally confirm that smaller, older and more polluting units are more likely candidates for retirement. In the table above, the second column gives the estimated value of the coefficient, which describes the relationship between the independent variables and the likelihood of retirement. The remaining columns give the standard error and the significance level. For example, the coefficient on net summer capacity is negative, which means that an increase in capacity decreases the probability that a unit is planned for retirement. Furthermore, as shown in table 2, the estimated coefficient is significant at the 6 percent level. An estimated coefficient is typically considered statistically significant if the significance is less than 10 percent and very significant if it is less than 5 percent. Similarly, the coefficient on unit age is positive, which means that an older unit is more likely to be retired, and this coefficient estimate is significant at the 1 percent level. The coefficients on SO_2 and NO_x emissions are also positive and significant at the 1 percent level.

Using the resulting logit regression equation, we analyzed "marginal effects" of changes in each of the independent variables on plans to retire an "average" unit owned by a power company in (1) a traditionally regulated market and (2) a restructured market, and the "average unit," for this purpose, is one with median values for age, size/net summer capacity, SO_2, and NO_x emissions rates, as shown in tables 3 and 4.

Table 3: Effect of 10% Change in a Variable Value on the Probability of an Average Unit's Planned Retirement When Owned by a Power Company in a Restructured Marked

Variable name	Median value of variable	Value with 10% change	Change in probability of retirement
Net summer capacity (size of unit)	193	212	-2%
Age of unit	43	47	7%
Unit's SO_2 emissions rate	0.49	0.54	1%
Unit's NO_x emissions rate	0.20	0.22	2%

Source: GAO analysis of Ventyx data.

Table 4: Effect of 10% Change in a Variable Value on the Probability of an Average Unit's Planned Retirement When Owned by a Traditionally Regulated Power Company

Variable name	Median value of variable	Value with 10% change	Change in probability of retirement
Net summer capacity (size of unit)	230	253	-1%
Age of unit	43	47	4%
Unit's SO_2 emissions rate	0.57	0.63	1%
Unit's NO_x emissions rate	0.23	0.25	1%

Source: GAO analysis of Ventyx data.

For example, a 10 percent increase in the capacity of an average unit owned by a power company in a restructured market, from 193 to 212 MW, would decrease the probability of that unit's retirement by about 2 percent, all other variables being held constant. For a unit owned by a power company in a traditionally regulated market, the same 10 percent would decrease the probability of retirement by about 1 percent. Note that the median values for units owned by power companies operating in traditionally regulated and restructured markets are not the same and that a 10 percent increase is therefore different.[7]

Analysis Indicates Units Power Companies Likely to Consider Retiring

The next step in our analysis was to use the resulting logit regression equation to estimate the number and generating capacity of other coal-fueled units that companies are likely to consider retiring among units belonging to companies that have not, as of yet, announced plans to retire coal-fueled units. We also estimated the generation associated with these potential retirements in megawatt-hours (MWh). We assume that some or all of these companies are likely to retire coal-fueled units, but that they either have not decided which ones, or simply have not publicly announced their plans. We further assume that these companies have or will base their decisions on the same characteristics as the companies that have already made announcements. Table 5 shows our analysis of units that power companies may consider for retirement by 2020.

---

[7]Note that marginal effects must be calculated for a given unit. We choose the median values for each variable, but other values could have been chosen. Unlike a linear regression analysis, the marginal effect of a logistic regression analysis depends on the value of all of the variables. This is a general feature and prevents the predicted probabilities from being greater than one or less than zero.

Table 5: Coal-Fueled Units That Companies May Consider for Retirement by 2020

	Units owned by companies that have announced at least one coal-fueled unit retirement	Units owned by companies that have announced *no* coal-fueled unit retirements	Total coal-fueled units
All units			
Number	482	477	959
Net summer capacity	158,000 MW	148,000 MW	306,000 MW
Units planned for retirement or that may be considered for retirement			
Number	174	90 to 138	264 to 312
Net summer capacity	30,400 MW	15,700 to 25,200 MW	46,100 to 55,600 MW
Net summer capacity as percent of all units	19%	11 to 17%	15% to 18%
Generation	150 million MWh	91 to 151 million MWh	241 to 301 million MWh
Generation as a percent of all units	16%	10 to 16%	13 to 16%

Source: GAO analysis of Ventyx data .

Note: We include only units from the Ventyx databases that have reported electricity generation and SO_2 emissions greater than zero during any of the years 2008 through 2012, and with net summer generating capacity greater than 25 MW. We rounded capacity numbers to the nearest 100 MW and generation to the nearest million MWh. Generation is based on annual average generation 2007-2011.

As shown in table 5, for the group of coal-fueled units whose owners have not reported any coal-fueled unit retirements, our analysis indicates from 90 to 138 units may likely be considered for retirement by 2020. This range represents the 95 percent confidence interval around our point estimate of 114 units. In other words, our model indicates that there is a 95 percent probability that the actual number of units that will retire is within this range. These 90 to 138 units account for 15,700 to 25,200 MW of capacity and 91 to 151 million MWh of electricity generation. If we add these units to those that power companies have announced for retirement, the total of coal-fueled retirements could range from 264 to 312 units by 2020, amounting to from 46,100 to 55,600 MW of capacity and average annual generation of 241 to 301 million MWh.[8] In percentage terms, this would be 15 to 18 percent of the capacity and 13 to 16 percent of the generation of the current coal-fueled fleet of generating units.

Limitations and Alternative Model Specifications

This section discusses the limitations of our model and alternative model specifications that we tested.

Limitations

A major limitation of our model is that we used a nonrandom sample of the entire population of coal-fueled units to estimate the relationship between the characteristics of coal-fueled units and power companies' plans to retire a unit. Our sample consisted of companies that announced plans to retire at least one unit but was not a random sample. It is possible that the companies that announced planned retirements and those that did not so announce differ in systematic ways that we do not observe from the data.[9] Such differences could result in omitted variable bias.

Another important limitation of our model is that we did not include all factors that contribute to power companies' decision to retire coal-fueled units. Apart from unit-level considerations, major factors that affect a power company's decision to retire a coal-fueled unit include fuel costs, environmental regulations, regional and local market considerations (e.g., expected future electricity demand and supply conditions, and transmission constraints), and technological developments in electricity generation and pollution control. For example, we did not take into account that planned unit retirements might make otherwise marginal

---

[8]We base our estimate of the expected loss of generation due to these retirements on generation from units in our dataset for the years 2007-2011.

[9]In addition, we could not distinguish between two types of power companies with no announced planned retirements: (1) on the one hand, companies that may actually have some plans to retire some of their coal-fueled units but, for one reason or another, have no incentive to make their plans public and (2) companies that have actually decided not to retire any of their units. In our dataset, both companies would be labeled "unannounced."

units in some regions more valuable and less likely to retire. Companies that own coal-fueled units may have different expectations regarding these factors, which we did not consider in our analysis.[10] Effectively, therefore, we assumed that power companies have very similar expectations regarding these factors.

These above limitations could mean that our model does not accurately or fully reflect power companies' unit retirement decisions. This would also mean that our estimates of how many unannounced units will retire may be inaccurate. For most of the limitations, the direction of bias in our model—the extent to which it may over- or under-estimate the likelihood of a unit retiring—is unclear. Addressing these limitations was beyond the scope of our review.

Alternative Specifications

To check the robustness of our model, we tested different specifications; that is, we ran logistic regressions using different sets of independent variables. For example, we tried specifications that included a measure of a unit's fuel efficiency, and another representing whether a unit is planning to install pollution control equipment. We also tried a version with unit average capacity factors in recent years, a measure of how intensively a unit is utilized. Based on our results, none of these variables significantly improved the model. Below, we discuss two other alternative specifications in more detail.

In one alternative specification, we used clustered standard errors. Our model assumes that each individual coal-fueled unit has a unique error term that is independent of every other unit. In this specification, we allow for the possibility that units owned by the same power companies may be related in unobserved ways and, therefore, the error terms may be correlated. As shown in table 6, the estimated coefficients in this alternative specification are very similar to our model, but the standard errors are generally bigger, and the estimated coefficients are generally less statistically significant. This is especially true for net summer capacity, which is no longer statistically significant at the commonly accepted 10 percent level.

---

[10]To do so would have required an extensive survey of power companies, which was beyond the scope of our review.

Table 6: Regression Results of Characteristics Associated with Planned Coal-Fueled Unit Retirements Using Clustered Standard Errors

Variable	Coefficient	Standard error	Significance[a]
Net summer capacity	-0.00198	0.0014	14.9%
Unit age	0.06647	0.0269	1.3%
SO_2 emissions rate	0.68053	0.2041	0.1%
NO_x emissions rate	3.64861	1.1659	0.2%
Regulatory status	-1.56564	0.3039	0.0%
Constant	-3.84944	1.7726	3.0%

Source: GAO analysis of Ventyx data.

Note: Number of observations equals 482.

[a]Significance is measured by the probability that we would find these results if the true value of the coefficient were zero. A lower significance level means that it is less likely that the observed pattern in the data is the product of pure chance. As with any regression, the statistical significance depends on the model being correctly specified.

In a second alternative specification, we used adjusted marginal cost as a proxy for the profitability of a unit. Based on economic logic and what we heard from stakeholders, we expected some indicator of the cost and profitability of electricity generation to contribute significantly to the retirement decision. Table 7 shows a version with marginal cost adjusted for regional wholesale prices and an interaction term with marginal cost and regulatory status. We adjusted marginal cost by dividing it by the regional wholesale price to account for the fact that units are more or less valuable depending on regional wholesale electricity prices. The interaction term allows us to effectively estimate two coefficients for adjusted marginal cost, one for power companies in traditionally regulated markets, and one for power companies in restructured markets. We included an interaction term to account for the possibility that power companies in traditionally regulated and restructured markets view costs differently.[11] Indeed, as shown in table 7, the estimated adjusted marginal cost coefficients differ—for power companies in restructured markets, the adjusted marginal cost coefficient is about 5.8, while the estimated coefficient for power companies in traditionally regulated markets is the adjusted marginal cost coefficient plus the interaction term (or 5.8 plus -

---

[11]State public utility commissions review the costs of traditionally regulated power companies and also review the prices that they charge for the electricity that they sell to their customers, while companies that operate under restructured markets are more directly exposed to market conditions governing costs and prices.

8.2 = -2.4). These results suggest that while higher adjusted marginal costs increase the probability of retirement of units owned by power companies in restructured markets, they decrease the probability for units owned by traditionally regulated power companies. The interpretation of these results is unclear.

Table 7: Regression Results of Characteristics Associated with Planned Coal-Fueled Unit Retirements Including Adjusted Marginal Cost and Interaction Term

Variable	Coefficient	Standard error	Significance[a]
Net summer capacity	-0.0019	0.0011	8.9%
Unit age	0.0667	0.0204	0.1%
SO_2 emissions rate	0.8113	0.1763	0.0%
NO_x emissions rate	3.4200	0.9328	0.0%
Adjusted marginal cost	5.8028	2.6250	2.7%
Cost and regulation interaction	-8.1719	2.7112	0.3%
Regulatory status	3.1255	1.4965	3.7%
Constant	-6.9313	2.0123	0.1%

Source: GAO analysis of Ventyx, FERC, and ICE data.

Note: Number of observations equals 482.

[a]Significance is measured by the probability that we would find these results if the true value of the coefficient were zero. A lower significance level means that it is less likely that the observed pattern in the data is the product of pure chance. As with any regression, the statistical significance depends on the model being correctly specified.

Regarding the costs of producing electricity, our findings differed for companies in restructured markets and companies that are traditionally regulated. Specifically, our results suggest that companies in restructured markets are more likely to retire units with higher adjusted marginal costs. In contrast, our results suggest that companies operating in regulated markets are less likely to retire units with higher adjusted marginal costs. A number of characteristics, not considered in our model, could provide alternative explanations for this difference. For example, it could be the case that the units in our sample have unique characteristics. One such potential case could be that units owned by power companies in traditionally regulated markets may be located in areas where concerns about the reliability of the electricity system are significant, and the costs of retrofitting an older generating unit are less costly than retiring it. Similarly, it could be that our sample contains a number of units located in areas with lower cost alternative suppliers or where prices are low—diminishing the attractiveness of even a relatively low-cost unit.

Appendix II: Description of Selected Scenarios and Forecasts

Table 8 describes key scenarios and assumptions in the EIA, IEA, and IHS Global Insight forecasts discussed in this report.

Table 8: Description of EIA, IHS Global Insight, and IEA Scenarios and Forecasts

Scenario	Description	Average annual change in electricity produced from coal, 2010-2035 (percentage)[a]
EIA Annual Energy Outlook		
Reference	A business-as-usual trend estimate, given known technology and technological and demographic trends. Baseline economic growth (2.5 percent per year from 2010 through 2035), oil price, and technology assumptions.	0.1%
Low Coal Price	Regional productivity growth rates for coal mining are approximately 2.8 percent per year higher than in the Reference scenario, and coal mining wages, mine equipment, and coal transportation rates in 2035 are between 21 and 25 percent lower than in the Reference scenario. EIA refers to this as its "Low Coal Cost" scenario.	0.5%
High Coal Price	Regional productivity growth rates for coal mining are approximately 2.8 percent per year lower than in the Reference scenario, and coal mining wages, mine equipment, and coal transportation rates in 2035 are between 25 and 27 percent higher than in the Reference scenario. EIA refers to this as its "High Coal Cost" scenario.	-0.6%
Low Gas Price	Assumes the well spacing for all tight oil and shale gas plays is 8 wells per square mile (i.e., each well has an average drainage area of 80 acres) and assumes 50 percent higher recovery for tight oil and shale gas wells than in the Reference scenario. Also assumes higher reserves of tight oil and shale gas that are more than twice the assumptions in the Reference scenario. EIA refers to this as its "High Technically Recoverable Resources" scenario.	-.7%
High Gas Price	Assumes 50 percent lower recovery per tight oil and shale gas well than in the Reference scenario, increasing the per-unit cost of developing the resource. Also assumes about 50 percent lower reserves of tight oil and shale gas than in the Reference scenario. EIA refers to this as its "Low Estimated Ultimate Recovery" scenario.	0.3%
$15 Greenhouse Gas Price	Applies a price for carbon dioxide emissions throughout the economy, starting at $15 per metric ton in 2013 and rising by 5 percent per year through 2035. In this scenario, carbon dioxide emissions from the electric power sector in 2035 would be reduced by 46 percent from 2010 levels.	-3.5%
$25 Greenhouse Gas Price	Applies a price for carbon dioxide emissions throughout the economy, starting at $25 per metric ton in 2013 and rising by 5 percent per year through 2035. In this scenario, carbon dioxide emissions from the electric power sector in 2035 would be reduced by 76 percent from 2010 levels.	-9.2%

Scenario	Description	Average annual change in electricity produced from coal, 2010-2035 (percentage)[a]
High Economic Growth	Assumes higher growth rate of gross domestic product—an average annual rate of 3 percent from 2010 to 2035, compared with 2.5 percent in the reference case.	0.3%
High Use of Energy Efficient Technologies	Assumes all future residential and commercial equipment purchases represent the most efficient models available in a particular year for each fuel, regardless of cost. EIA refers to this as its "Integrated Best Available Demand Technology" scenario.	-1.1%
IHS Global Insight		
	A long-term forecast based on a single set of assumptions about future energy policies, fuel prices, and other factors. For example, IHS Global Insight assumes that a federal greenhouse gas cap-and-trade program is implemented for the power sector in 2021 but with a limit on how high carbon prices could rise; a federal clean energy standard is enacted to promote use of electricity generated using renewables, nuclear energy, and coal technologies capable of carbon-capture; and improvements are made in the energy efficiency of appliances, equipment and building standards.	-0.5%
IEA World Energy Outlook[a]		
Current policies	A baseline scenario based on government policies and measures enacted by mid-2011.	0.8%
New policies	Assumes some future implementation of international policy commitments and plans related to climate change, pollution, and other energy-related challenges. This includes, for example, implementation of EPA regulations to reduce mercury and other pollutants in the power sector.	-0.1%
Scenario 3	Assumes future changes consistent with a 17 percent reduction in U.S. carbon dioxide emissions in 2020 compared with 2005, as well as the implementation of a price on carbon dioxide after 2020 in the United States. This scenario represents a reduction of U.S. carbon dioxide emissions from energy of more than half by 2035 compared with emissions in 2009.	-3.0%

Sources: EIA, IHS Global Insight, IEA.

[a]IEA data reflect change from 2009 to 2035.

Appendix III: GAO Contact and Staff Acknowledgments

GAO Contact	Frank Rusco, (202) 512-3841, ruscof@gao.gov
Staff Acknowledgments	In addition to the contact named above, Jon Ludwigson (Assistant Director), Mike Armes, Patrick Dudley, Philip Farah, Quindi Franco, Cindy Gilbert, Paige Gilbreath, Alison O'Neill, Kendal Robinson, Jeanette Soares, and Kiki Theodoropolous made key contributions to this report.

Related GAO Products

EPA Regulations and Electricity: Better Monitoring by Agencies Could Strengthen Efforts to Address Potential Challenges. GAO-12-635. Washington, D.C.: July 17, 2012.

Air Emissions and Electricity Generation at U.S. Power Plants. GAO-12-545R. Washington, D.C.: April 18, 2012.

Coal Power Plants: Opportunities Exist for DOE to Provide Better Information on the Maturity of Key Technologies to Reduce Carbon Dioxide Emissions. GAO-10-675. Washington, D.C.: June 16, 2010.

Clean Coal: DOE's Decision to Restructure FutureGen Should Be Based on a Comprehensive Analysis of Costs, Benefits, and Risks. GAO-09-248. Washington, D.C.: February 13, 2009.

Climate Change: Federal Actions Will Greatly Affect the Viability of Carbon Capture and Storage As a Key Mitigation Option. GAO-08-1080. Washington, D.C.: September 30, 2008.

Restructured Electricity Markets: Three States' Experiences in Adding Generating Capacity. GAO-02-427. Washington, D.C.: May 24, 2002.